AMAZING BUILDINGS

BLACK RABBIT BOOKS

JOANNE MATTERN

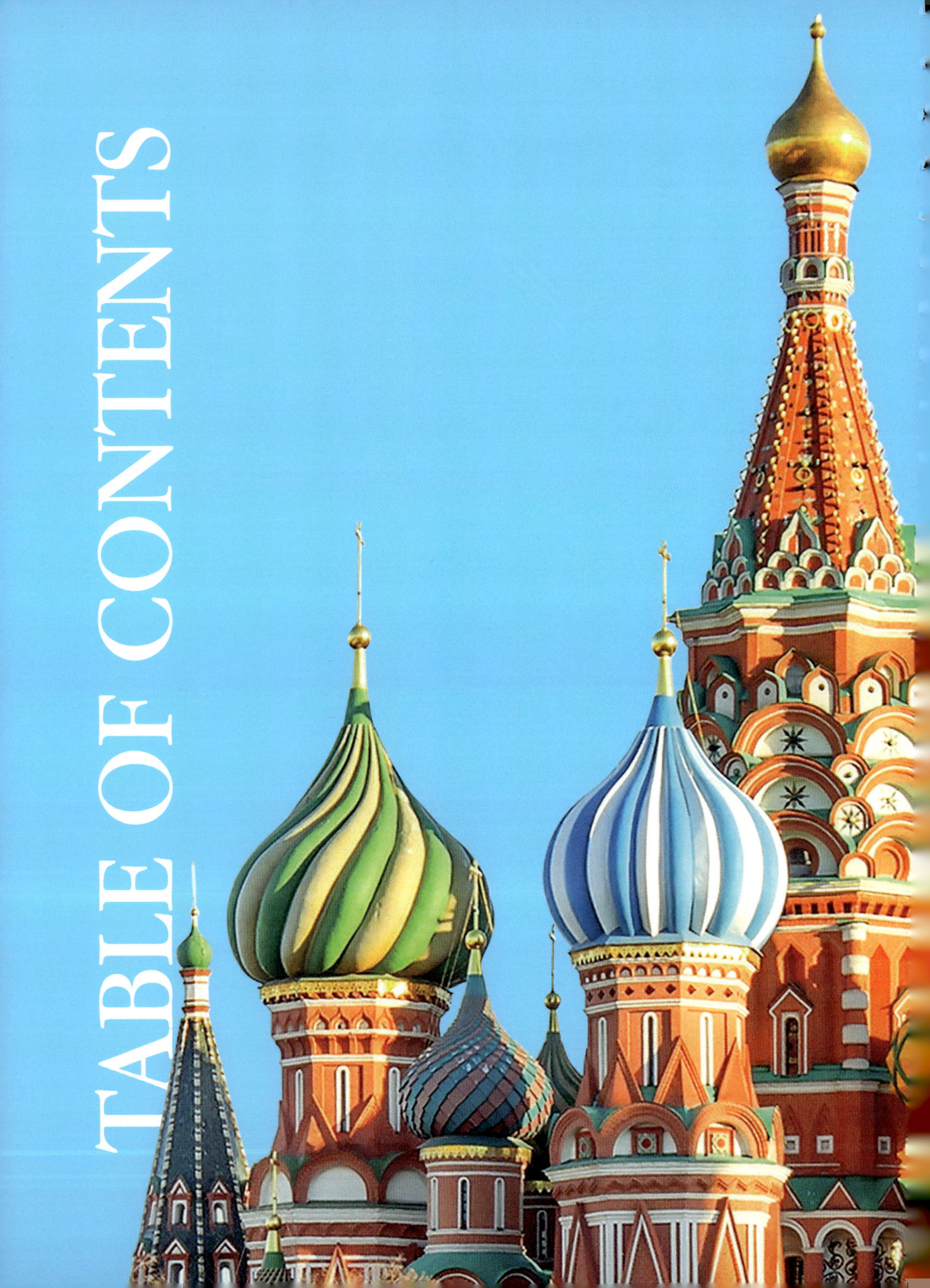
TABLE OF CONTENTS

Cambodia's
Angkor Wat

Angkor Wat (ANG-kawr WAHT) is one of the largest religious monuments. It covers over 154 square miles (400 sq kilometers). It was built by a king. This was in the 1100s. The name means "City of Temples." It was built with sandstone blocks.

The temple honored kings and gods. First, it was a Hindu temple. Next, it became a Buddhist temple. There were many wars in the 1400s. It was abandoned then.

Today, the temple is popular with tourists. More than 2 million visit each year.

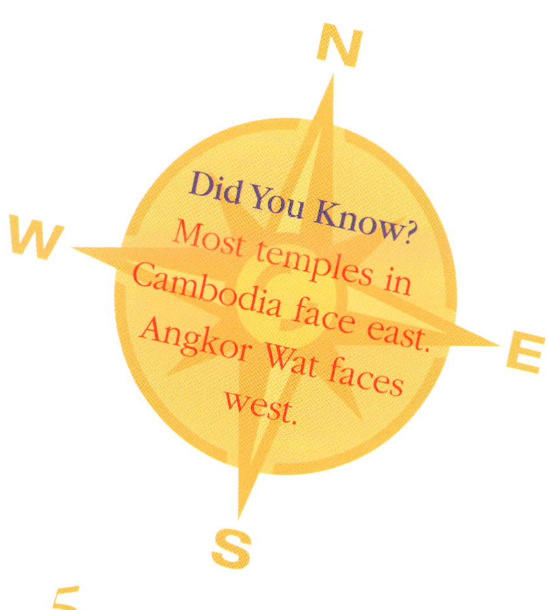

Dubai's Burj

Burj Khalifa (BERJ kuh-LEE-fuh) is the tallest building in the world. It is in Dubai. It soars high into the sky. The tower can be seen from 60 miles (95 km) away.

The Burj was built in 2010. It is 2,717 feet (828 meters) tall. It has 209 floors. The top 46 floors are just for structure. People do live in the Burj. It also has offices and restaurants. There are three **observation decks**. The highest is on the 148th floor. The views are incredible.

Khalifa

Did You Know?
Burj Khalifa will not be the tallest building for long.
Another tower is rising in Dubai. It will be even taller.

View from the 124th floor observation deck.

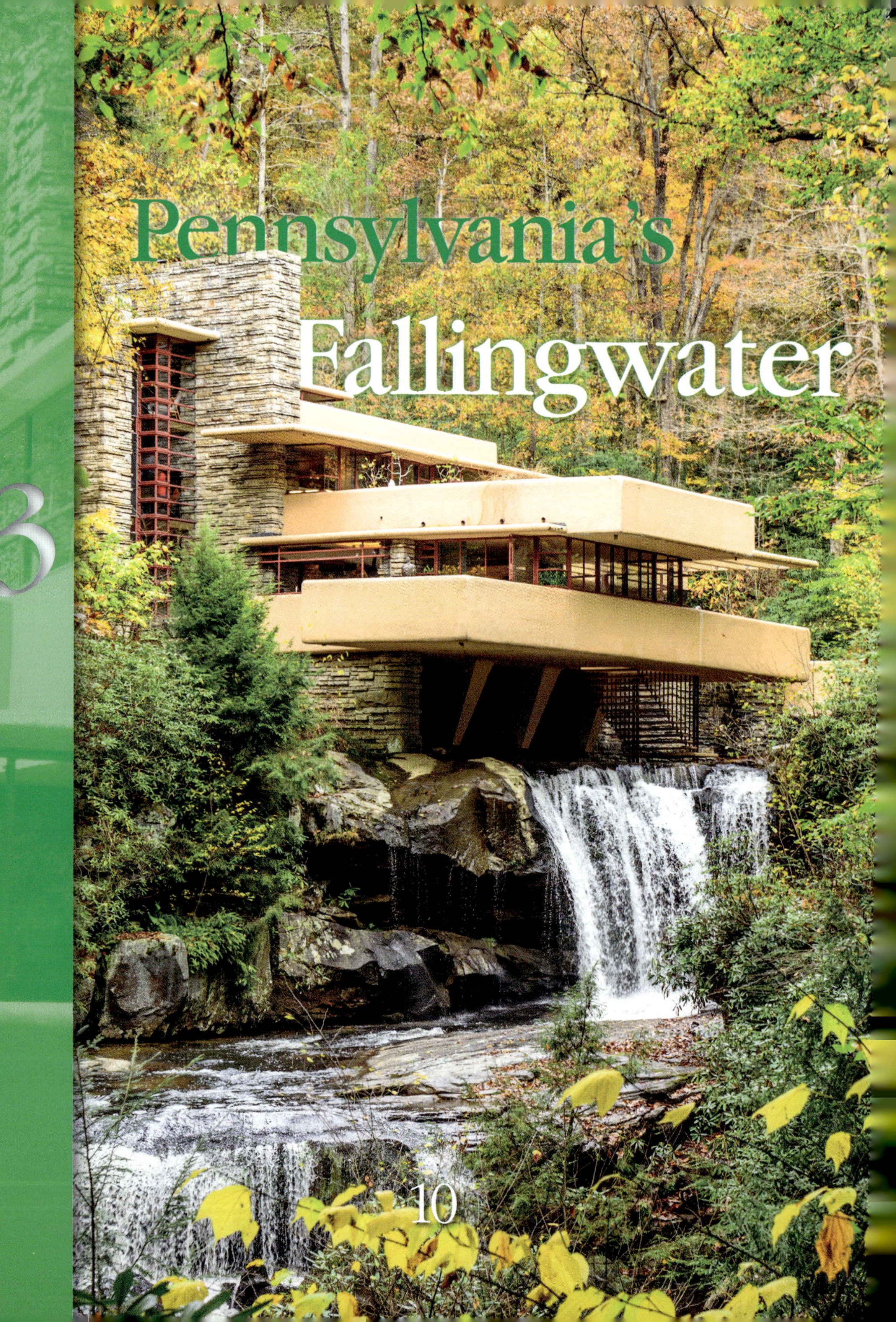

3
Pennsylvania's
Fallingwater
10

Frank Lloyd Wright was an American **architect**. He believed buildings should blend in with nature. Fallingwater is one of his most beautiful designs. It was a private home. It seems to flow into the trees and rocks around it.

The owners wanted a view of a waterfall. Instead, Wright put their house on top of the waterfall. The water became part of his design. The house was completed in 1938. It is 5,330 square feet (495 sq m).

Frank Lloyd Wright

11

Think
About It

Should
buildings
blend
in with
nature?
Why or
why not?

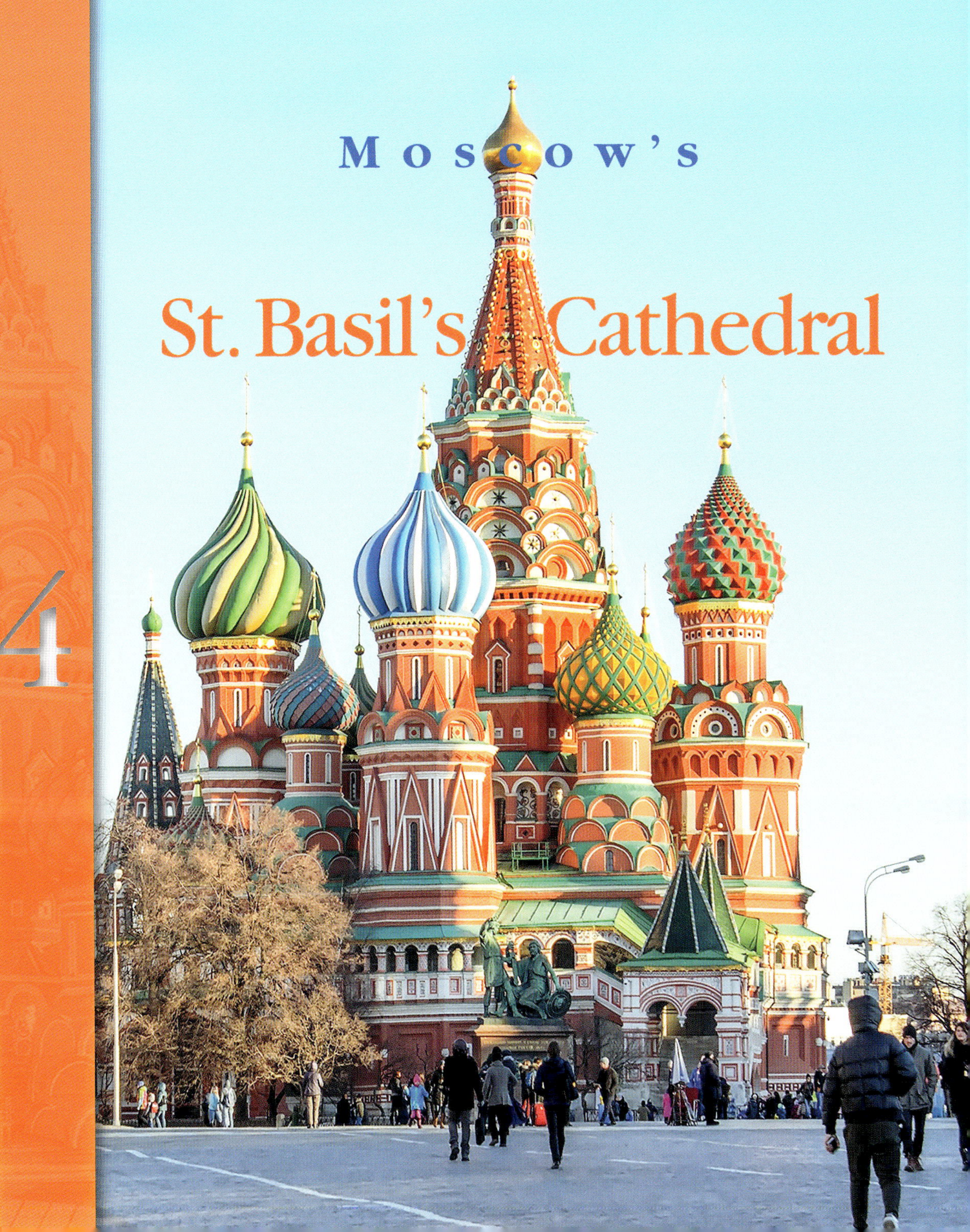

Moscow's

St. Basil's Cathedral

Ivan the Terrible was the first **czar** of Russia. He built
St. Basil's **Cathedral**. It celebrated his **military** victories.
The church was finished in 1561. It is in the center of a
public area called Red Square.

The cathedral has expanded over the
years. It wasn't always so colorful.
At first, it was white with gold domes.
Colors were added in the 1680s. Today's
color scheme dates from 1860. There are
nine chapels. One original bell remains
from the cathedral. It is still used today.

*Ivan the Terrible
monument.*

Did You Know? St. Basil's was a museum. In 1991 it started holding religious services again.

5
Prague's
Dancing
House

Instead of standing straight and tall, the Dancing House is curved. The towers in front look like a couple dancing. Sometimes the house is called the Fred and Ginger Building. That name comes from Fred Astaire and Ginger Rogers. They were Hollywood dancers from the 1930s.

The house was finished in 1996. It is a 40-room hotel and restaurant. At first, people didn't like it. Today, it is one of the most popular places to visit. It has beautiful views of the city.

Fred Astaire and Ginger Rogers.

Think About It
Why didn't people like the Dancing House at first?   Why do they like it now?
18

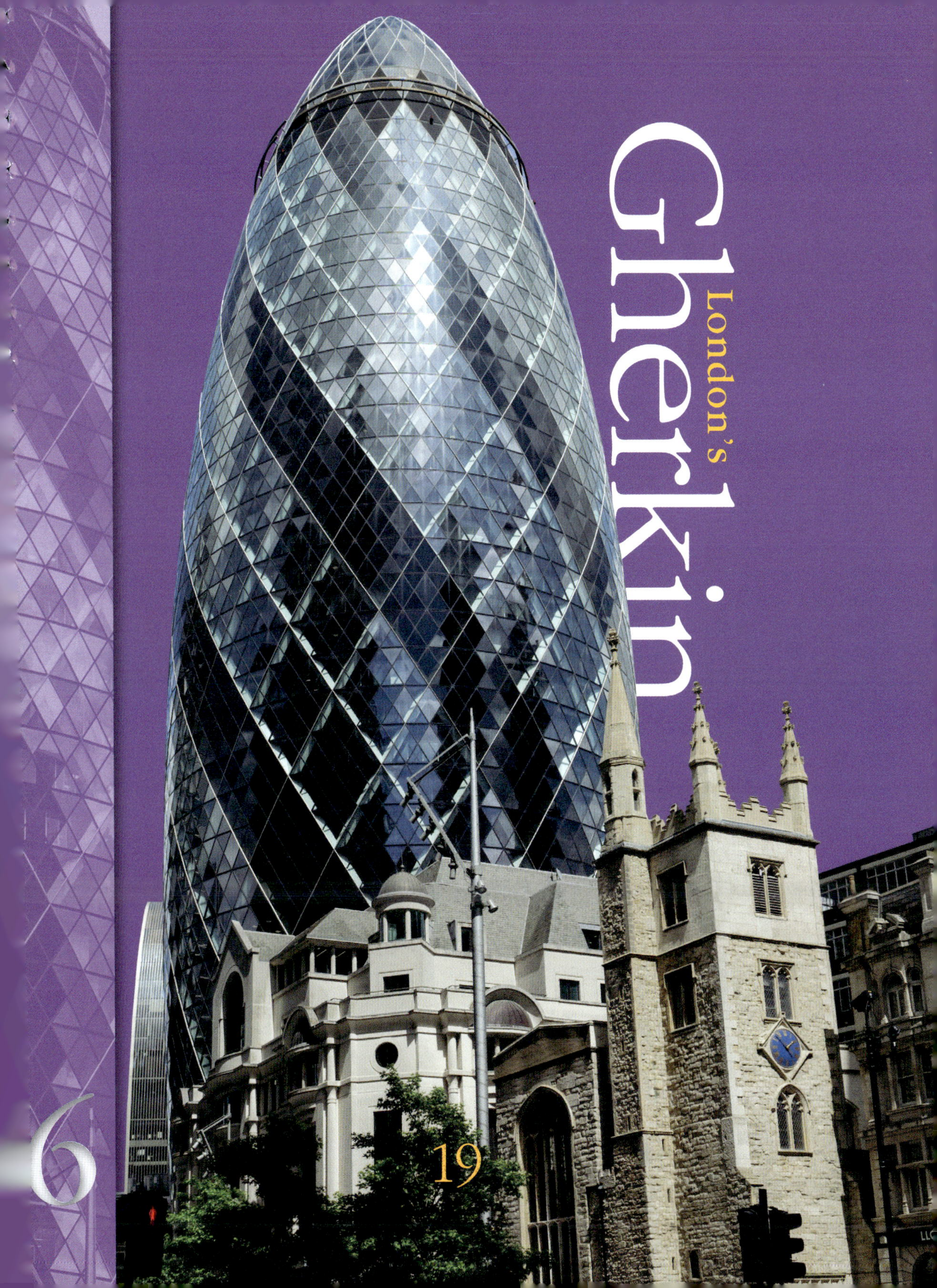

London's
Gherkin
6
19

A gherkin (guhr-KIN) is a kind of pickle. It is also a building in London. The Gherkin's real name is 30 St. Mary Axe. It looks like a tall, fat pickle.

The Gherkin opened in 2004. It is an office building. It has almost 260,000 square feet (24,000 sq m) of glass. There is a big piece of curved glass at the top.

All that glass catches the light. The light spreads through the building. The Gherkin uses half the energy of other buildings its size.

Did You Know?

The distance around the Gherkin is almost equal to its height.

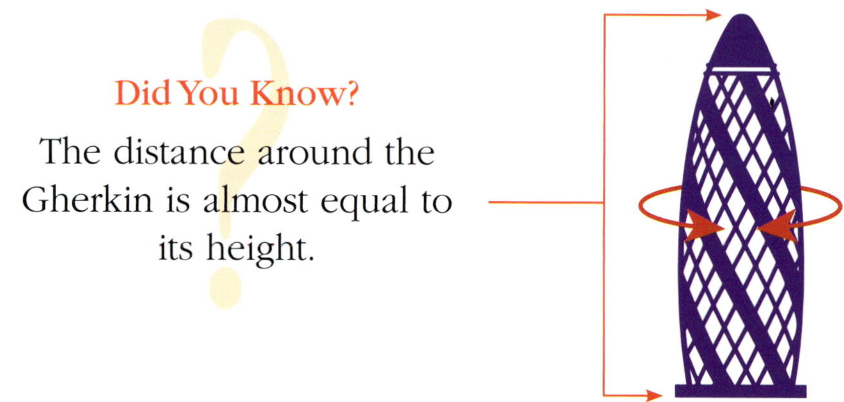

MORE TO EXPLORE

WHERE IN THE WORLD

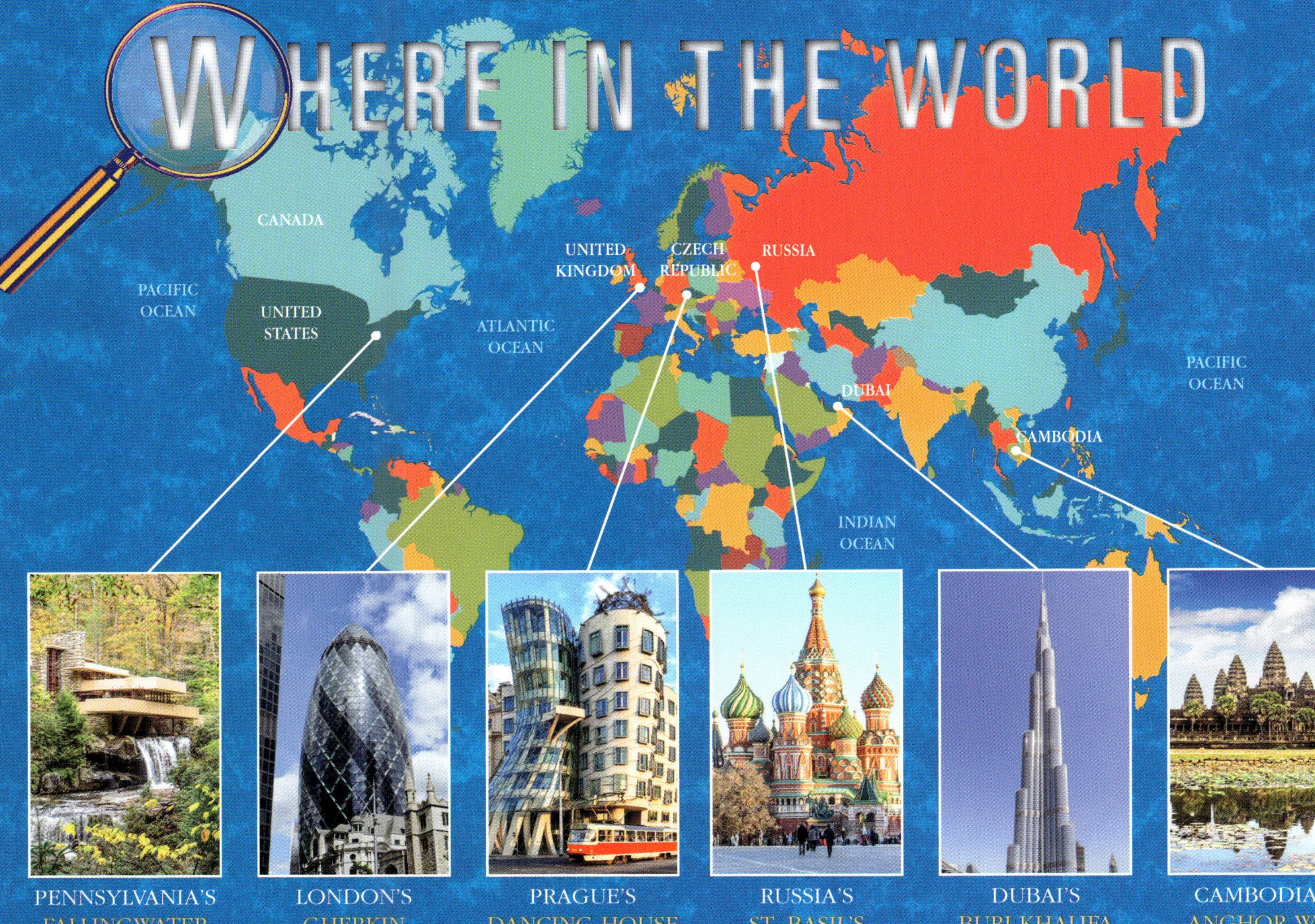

PENNSYLVANIA'S
FALLINGWATER

LONDON'S
GHERKIN

PRAGUE'S
DANCING HOUSE

RUSSIA'S
ST. BASIL'S
CATHEDRAL

DUBAI'S
BURJ KHALIFA

CAMBODIA'S
ANGHOR WAT

FANTASTIC FACTS

Angkor Wat is shown on the Cambodian flag. It is also on some Cambodian money.

Burj Khalifa's elevator travels farther than any other building elevator.

Frank Lloyd Wright only used two colors of paint in Fallingwater. They are light ochre and red.

Ivan the Terrible had a bad temper. According to legend, he blinded St. Basil's architect. That way, he could never copy the design for another building.

The Dancing House is made of 99 concrete panels. They are all different shapes and sizes.

When the Gherkin was being built, the ancient skeleton of a Roman girl was found. She was later reburied at its base.

COOL COMPARISONS

How do these amazing buildings measure up?

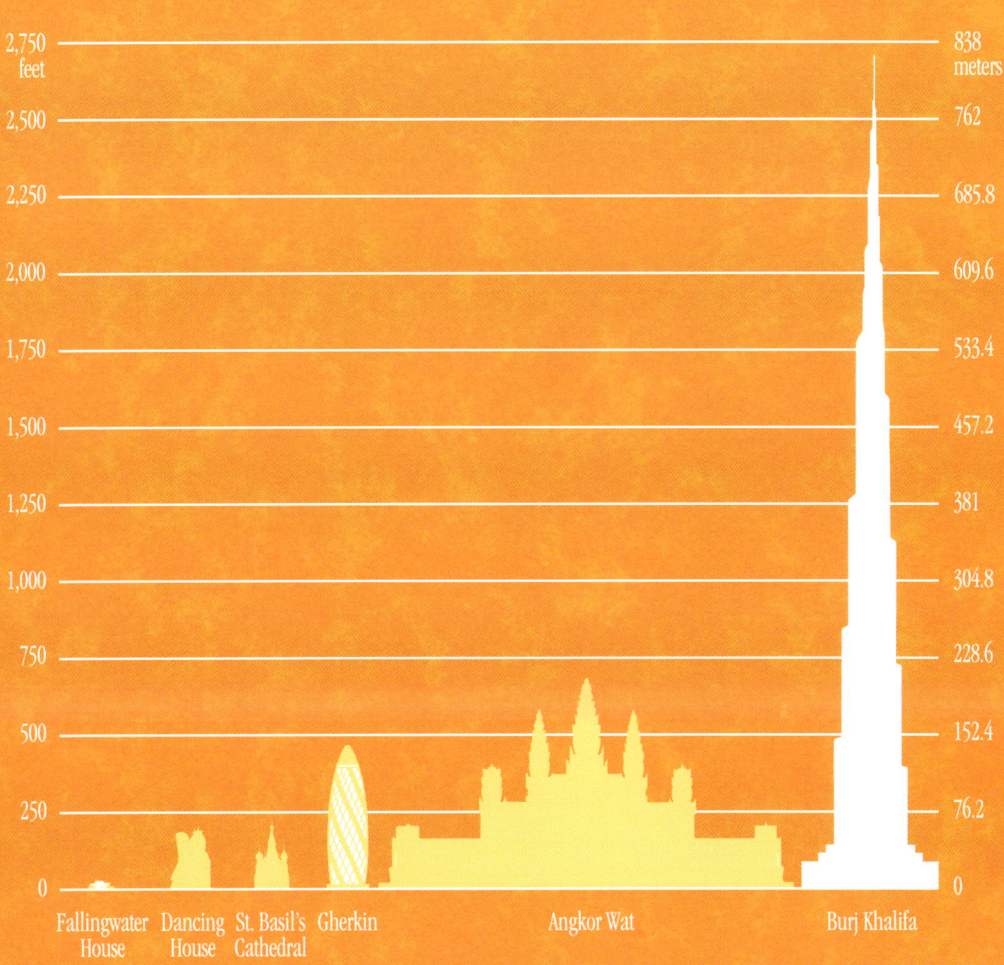

RESOURCES

Glossary

architect (AHR-kuh-tekt) A person who designs buildings.

cathedral (kuh-THEE-druhl) A large and important church.

czar (ZAHR) An emperor of Russia.

military (MIL-uh-ter-ee) Having to do with the armed forces.

monument (MON-yuh-muhnt) A building or statue that honors a person or event.

observation deck (ob-zur-VAY-shuhn DEK) An area in a tall building where people can look out at the view.

Read More

Dittmer, Lori. *Empire State Building*. Mankato, MN: Creative Education/Creative Paperbacks, 2020.

Duling, Kaitlyn. *Safe Buildings*. Minneapolis: Jump!, Inc., 2019.

Index

TOP RANK is published by Black Rabbit Books, P.O. Box 227, Mankato, MN, 56002. • COPYRIGHT © 2025 Black Rabbit Books. All rights reserved. No part of this book may be reproduced in any form without written permission from the publisher. • Top Rank is an imprint of Black Rabbit Books. • Edited by Alissa Thielges • Designed by Danny Nanos • Photographs © Flickr: Jeremy Thompson, 12; Getty: finallast, 15; Shutterstock: Alexey Borodin, 14, Ami Parikh, cover, 6–7, 21, domnitsky, 20, Jaguar PS, 17, Mary_Photo, 2–3, 13, 21, Michaela Jilkova, 18, Orion Media Group, 8–9, Mauro Gea, 19, 21, Sean Pavone, 10, 21, seeyah panwan, 21, Sergii Figurnyi, 4, 21, soponyono, 5, spatuletail, 11, Vlas Telino studio, 16, 21 • Printed in the United States of America

Library of Congress Cataloging-in-Publication Data: Names: Mattern, Joanne, 1963- author. Title: Amazing buildings / by Joanne Mattern. | Description: Mankato, MN: Top Rank, an imprint of Black Rabbit Books, [2025] | Series: Design marvels | Ages 8–11 | Grades 4–6 | Identifiers: LCCN 2023058204 | ISBN 9781632357885 (library binding) | ISBN 9781645820673 (ebook) | Subjects: LCSH: Buildings—Juvenile literature. | Architecture—Juvenile literature. | Classification: LCC TH149 .M387 2025 | DDC 720—dc23/eng/20240112 | LC record available at https://lccn.loc.gov/2023058204